EAST AND NORTH EASTERN STEAM

EAST AND NORTH EASTERN STEAM
FORMER LNER TERRITORY 1947-1958

The Railway Photographs of Andrew Grant Forsyth

BRIAN J. DICKSON

The History Press

Front cover: Saturday 31 March 1951. Ex-LNER Class B1 4-6-0 No. 61168 is seen at the head of an 'up' express at Doncaster station.

Back cover: Thursday 16 August 1948. At Marylebone Station, ex-LNER Class B1 4-6-0 No. 61169 is waiting to depart at the head of the 10.00a.m. working to Manchester London Road. Constructed at the Vulcan Foundry during 1947, she would be withdrawn from traffic in 1963.

Half-title page: Saturday 9 July 1955. Seen near Potters Bar at the head of an 'down' Pullman is the premier Class A4 locomotive, which entered service from Doncaster Works in September 1935. No. 60014 *Silver Link* would originally be numbered 2509 and painted in grey livery and have the distinction of hauling the first running of the 'Silver Jubilee' from London to Newcastle. Allocated to both King's Cross and Grantham sheds throughout her working life, she would be withdrawn from service in December 1962.

Title page: Saturday 25 August 1956. In Bridlington shed yard, ex-LNER Class V2 2-6-2 No. 60831 is awaiting the return of the special working she brought into Bridlington from Sutton in Ashfield earlier in the day. The excursion was organised for the staff of the hosiery manufacturer I & R Morley, which is clearly stated on the huge headboard the locomotive is carrying. Entering service from Darlington Works in 1938, she would be rebuilt with three separate cylinders during 1957 and be withdrawn from service in 1966. (AGF483)

First published 2021

The History Press
97 St George's Place,
Cheltenham GL50 3QB
www.thehistorypress.co.uk

British Library Cataloguing in Publication Data.
A catalogue record for this book is available from the British Library.

ISBN 978 0 7509 9854 3

Typesetting and origination by The History Press
Printed in Turkey by Imak

INTRODUCTION

Andrew Grant Forsyth was born in Barnet, north London, in 1923 and lived and worked in the same area all his life. A lifelong railway enthusiast, he was very much a follower of all things associated with the London and North Eastern Railway and the greater part of his photographic collection contains images relating to that company and its pre-grouping constituents. His railway photography started in 1947 with the use of a Kodak folding camera and an Agfa Record camera, using 2¼in x 3½in (6cm x 9cm) film. He moved to 35mm film in 1950 with a Kodak Retina camera, which was followed by a Leica, a Pentax S3 and finally a Nikon F90.

His catalogue of photographs shows images from all parts of the former LNER territory, from Newcastle in the north to Darlington, Doncaster, York, Nottingham, Norwich and numerous sites in the north London area. A visit to the remote East Anglian branches in 1952 is of particular interest.

The selection seen here shows a wide variety of locomotives from the former Hull and Barnsley Railway tanks to the former North Eastern Railway 'Q' Classes, Nigel Gresley designed Class A3 and A4 and the Arthur Peppercorn 'Pacifics' photographed between 1947 and 1958, when Andrew was called up for National Service. The greater part of this selection shows locomotives and workings in the former North Eastern Region of British Railways.

In his spare time Andrew was one of the organisers of the Seafield Railway Club, which arranged visits to sheds and railway sites throughout eastern England and Scotland. The club also published a regular magazine titled *The Locomotive Post*, containing observations from contributing enthusiasts and detailing locomotive stock changes and movements that were taking place throughout the industry.

Prior to his death at the age of 83, his photographic collection had already been carefully managed by Initial Photographics.

Grateful thanks are due to Driver John Webb for sharing his route knowledge of the former Great Northern line out of King's Cross.

Opposite top: Monday 2 June 1947. Plodding along on the 'up' slow line near Greenwood on the former GNR line into King's Cross, LNER Class O4 2-8-0 No. 3682 is at the head of a goods train. Constructed by Robert Stephenson & Co. during 1917 as part of an order for the ROD of this former GCR Class 8K design, she would be numbered 1664 by the ROD and be purchased by the LNER in 1928. Numbered 6548 by them, she would later become No. 3682 and finally No. 63682 with British Railways, being withdrawn during 1959.

Opposite bottom: Saturday 26 July 1947. Introduced by Alfred Hill for the GER during 1915, his Class L77 (LNER Class N7) 0-6-2 tanks continued construction during the post-grouping period with the final examples entering service in 1928. The example seen here standing in Stratford shed yard, No. 9662, was a product of Robert Stephenson & Co. in 1925 that would utilise a Belpaire firebox. She would be rebuilt during 1955 with a round-top firebox and be withdrawn from service four years later in 1959.

Above: Tuesday 19 August 1947. Standing in a row of out-shopped ex-works locomotives at Darlington Works, three-cylinder LNER Class Q7 0-8-0 No. 3470 had been constructed at the same works during 1924. Based on a design by Vincent Raven and introduced to the NER in 1919 as their Class T3, five examples were constructed that year with a further ten examples following post-grouping in 1924. Specifically designed to handle heavy mineral traffic, many spent their days working out of Tyne Dock shed. The whole class of fifteen examples was withdrawn from traffic during December 1962.

Tuesday 19 August 1947. At Darlington Works, ex-Hull and Barnsley Railway (H&BR) Class F3 (LNER Class N13) 0-6-2 tank No. 9113 has just completed an overhaul and is being readied to enter service. Constructed by Hawthorn, Leslie & Co. during 1913, she would be numbered 23 by the H&BR, becoming No. 2415 and later No. 9113 with the LNER. British Railways would number her 69113 and she would end her days allocated to Neville Hill shed in Leeds, being withdrawn from service in 1953.

Wednesday 20 August 1947. Bearing full green LNER lined livery, ex-NER Class E (LNER Class J71) 0-6-0 tank No. 8286 is acting as one of the York station pilots on this day. Constructed at Darlington Works in 1892, she would be withdrawn from service sixty years later in 1952.

Wednesday 20 August 1947. Standing at the head of a passenger train at York station is former NER Class R (LNER Class D20) 4-4-0 No. 2389. Designed for express passenger work and introduced during 1899, the example seen here was constructed at Gateshead Works in 1907 incorporating a saturated steam boiler. Rebuilt during 1914 with a superheated boiler, she would be withdrawn from service in 1954 numbered 62389 by British Railways.

Above: Thursday 21 August 1947. Seen in sparkling ex-works condition at Doncaster Works, LNER Class A3 4-6-2 No. 108 *Gay Crusader* has just emerged wearing its new green livery. Constructed at the same works in 1923 as a Class A1 locomotive, she would be rebuilt as a Class A3 in 1943, acquiring a double chimney in 1959 and trough-style smoke deflectors during 1961. Named after the racehorse that won the 1917 Derby, St Leger and 2000 Guinea races, she would be withdrawn from service in 1963.

Opposite top: Thursday 21 August 1947. At the south end of Doncaster station, ex-GNR Class C1 (LNER Class C1) 4-4-2 No. 2849 is waiting to depart at the head of a passenger working. Constructed at Doncaster Works during 1906 with a saturated steam boiler, she was rebuilt in 1923 with a superheating boiler. In the same year she was fitted with an experimental booster engine on the trailing axle and was subject to a number of test workings with the LNER dynamometer car. The booster engine would be removed in 1935 and she would be withdrawn from service less than a year after this photograph in July 1948.

Opposite bottom: Thursday 21 August 1947. At Doncaster Works, ex-GNR Class J22 (LNER Class J6) 0-6-0 No. 4271 is standing in the yard. Constructed at the same works in 1921, she would be withdrawn from service after only thirty-seven years during 1958.

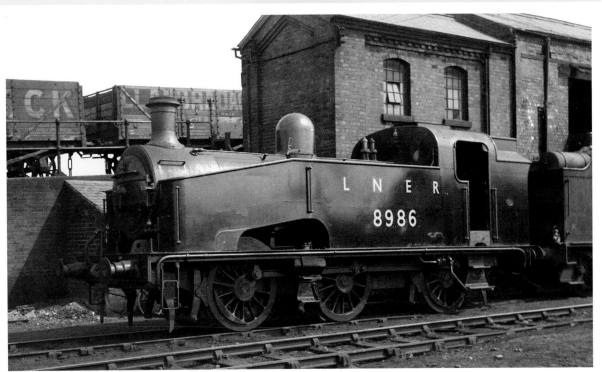

Thursday 21 August 1947. Standing adjacent to the coaling stage at Doncaster shed is LNER Class J50/4 0-6-0 tank No. 8986. Constructed at the former GCR Gorton Works during 1939, she would only give twenty-three years of service, being withdrawn in 1962.

Saturday 4 October 1947. Near Harringay station in north London, ex-GNR Class J14 (LNER Class J53) 0-6-0 saddle tank No. 8760 waits for the 'right of way' at the head of an 'up' goods train destined for west London. Constructed at Doncaster Works in 1893, she would be numbered 926 by the GNR, becoming No. 3926 and later No. 8760 with the LNER. British Railways would number her 68760, withdrawing her from service in 1954.

Saturday 25 October 1947. Standing in Hull Dairycoates shed yard is ex-NER Class B (LNER Class N8) 0-6-2 tank No. 9373. Entering service from Darlington Works in 1886, she would be numbered 535 by the NER, becoming No. 9373 with the LNER. She would be withdrawn from service during 1950 after sixty-four years of service.

Saturday 25 October 1947. Also seen in Hull Dairycoates shed yard is ex-NER Class P (LNER Class J24) 0-6-0 No. 5619. Entering service from Gateshead Works in 1896 and numbered 1860 by the NER, she would become No. 5619 with the LNER and No. 65619 with British Railways, being withdrawn from service during 1951. Note the unusual storage position for the firing shovel.

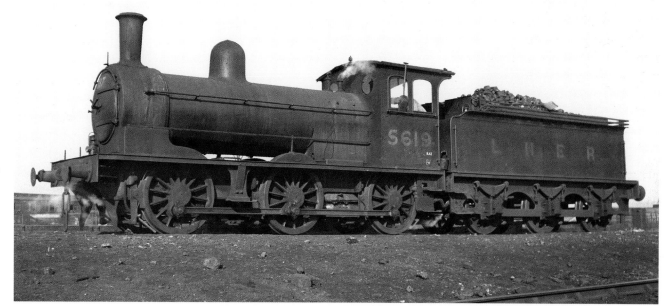

Opposite top: Saturday 25 October 1947. Designed as a heavy shunting locomotive by Wilson Worsdell, his three-cylinder Class X 4-8-0 for the NER comprised initially only ten examples coming out of Gateshead Works in 1909 and 1910. Five further examples were produced at Darlington Works in 1925 and all were designated Class T1 by the LNER. No. 9915 is seen here at Hull Dairycoates shed. Constructed in 1910, she would be withdrawn during 1959.

Opposite bottom: Saturday 13 March 1948. This diminutive but powerful shunting locomotive was one of only five members of the former GER Class B74 constructed at Stratford Works during 1914. Numbered 228 by the GER, she would become No. 7228 and later 8126 with the LNER. Designated Class Y4 by the LNER, this example seen here at Stratford shed would be withdrawn from service in 1957.

Above: Tuesday 30 March 1948. The photographer travelled west in England to reach an outpost of the former LNER working territory to visit the former Cheshire Lines Committee (CLC) shed at Brunswick near Liverpool central station, where ex-GCR Class C (LNER Class F1) 2-4-2 tank No. 7099 is seen in good clean condition. Constructed at Gorton Works during 1891, she would be withdrawn from service early in 1949.

Opposite top: Thursday 1 April 1948. Edward Thompson's only tank locomotive design for the LNER was his Class L1 2-6-4 tank, the first example appearing from Doncaster Works during 1945. It would be three years and much testing before any further examples entered service, with twenty-nine coming out of Darlington Works in 1948. Seen here at Darlington Bank Top station is No. ε9009. Having exited the works in the month prior to this photograph, she would be renumbered 67710 later in 1948 and be withdrawn after only fourteen years of service during 1962.

Opposite bottom: Friday 2 April 1948. At Hull Dairycoates shed, ex-GER Class M15 (LNER Class F4) 2-4-2 tank No. 7171 bears no owner's identity. Entering service from Stratford Works during 1908 and numbered 175 by the GER, she would become No. 7175 and later No. 7171 with the LNER. She would be withdrawn from service in 1951.

Above: Friday 2 April 1948. In a terribly grimy state with her former owner's name and number barely visible, ex-NER Class P1 (LNER Class J25) 0-6-0 No. 5666 is seen in Hull Dairycoates shed yard. Constructed at Gateshead Works during 1898, she would be numbered 1992 by the NER, becoming No. 5666 with the LNER. This locomotive was one of forty-two examples of the class that were loaned to the GWR during the Second World War. Becoming No. 65666 with British Railways, she would be withdrawn in 1960 after sixty-two years' service.

Opposite top: Friday 2 April 1948. In a similarly filthy condition, ex-NER Class T2 (LNER Class Q6) 0-8-0 No. 3382 is also seen in Hull Dairycoates shed yard. Constructed at Darlington Works in 1917, she would be numbered 2225 by the NER, becoming No. 3382 with the LNER. This numerous class designed by Vincent Raven and introduced in 1913 would total 120 examples, with the last entering service during 1921. Becoming No. 63382 with British Railways, she would be withdrawn from service in 1964.

Opposite bottom: Friday 2 April 1948. At Hull Botanic Gardens shed, ex-LNER Class B1 4-6-0 No. 1165 has yet to acquire her British Railways identity and livery. A product of the Vulcan Foundry during 1947, she would be allocated new to Gorton and end her days at Canklow, Rotherham, being withdrawn from service in 1964.

Above: Friday 2 April 1948. With 6ft 10in driving wheels, this three-cylinder ex-NER Class Z (LNER Class C7) 4-4-2 has a massive presence with its huge superheated boiler. Seen here at Hull Dairycoates shed, No. 2970 was the second example of the class to be constructed at Darlington Works in 1914. Numbered 2164 with the NER, she became No. 2970 with the LNER. She would be withdrawn eight months after this photograph in December 1948.

Opposite top: Friday 2 April 1948. The heavy three-cylinder 4-6-2 tank locomotives of the former NER Class Y (LNER Class A7) were primarily designed to assist in handling the large amounts of coal traffic coming out of the many collieries in their territory. A total of twenty examples were constructed at Darlington Works during 1910 and 1911, with No. 9770, seen here at Hull Dairycoates shed, being the first of the class to enter service in 1910. Numbered 1113 by the NER, she became No. 9770 with the LNER and No. 69770 with British Railways, being withdrawn during 1954.

Opposite bottom: Friday 2 April 1948. At Hull Botanic Gardens shed, ex-LNER Class D49 4-4-0 No. 2701 *Derbyshire* was the second member of this class to enter service from Darlington Works in 1927. She would be withdrawn from service thirty-two years later during 1959.

Above: Friday 2 April 1948. Also seen at Hull Botanic Gardens shed is the massive bulk of ex-NER Class D (LNER Class A8) 4-6-2 tank No. 9866. Originally constructed by the NER as a Class H1 4-4-4 tank at Darlington Works in 1914, she was rebuilt as a 4-6-2 during 1934 and withdrawn from service in 1958.

Opposite top: Saturday 3 April 1948. At the former GNR shed at Copley Hill in Leeds, ex-GNR Class N1 (LNER Class N1) 0-6-2 tank No. 9436 is parked within a row of locomotives. Designed primarily to handle the suburban traffic out of King's Cross in London and that in West Yorkshire, many examples would be based at Copley Hill. No. 9436 was constructed at Doncaster Works during 1907 and would be numbered 1556 by the GNR, becoming No. 4556 and later No. 9436 with the LNER. She would be withdrawn from service during 1955.

Opposite bottom: Saturday 3 April 1948. Sporting her new number and owner's identity, ex-NER Class C (LNER Class J21) 0-6-0 No. E5077 is seen standing in the yard at Leeds Neville Hill shed. Designed by Thomas Worsdell and introduced during 1886, construction of the class continued until 1895, with a total of 201 examples entering service. No. E5077 was constructed at Gateshead Works in 1891 as a two-cylinder compound locomotive with a saturated steam boiler. Rebuilt in 1910 as a two-cylinder simple and fitted with a superheating boiler during 1923, she would be withdrawn from service in 1953.

Above: Saturday 24 April 1948. The ex-GCR Class 1B (LNER Class L1 later L3) 2-6-4 tanks were originally designed to handle coal traffic from the Nottinghamshire and Derbyshire coal pits to Immingham, but found much use throughout the former GCR on general goods traffic. The massive bulk of No. 69054 is seen here at Neasden shed. Constructed at Gorton Works in 1915, she would be withdrawn from service during 1949. It should be noted that this class saw the first use in the UK of the 2-6-4 wheel arrangement.

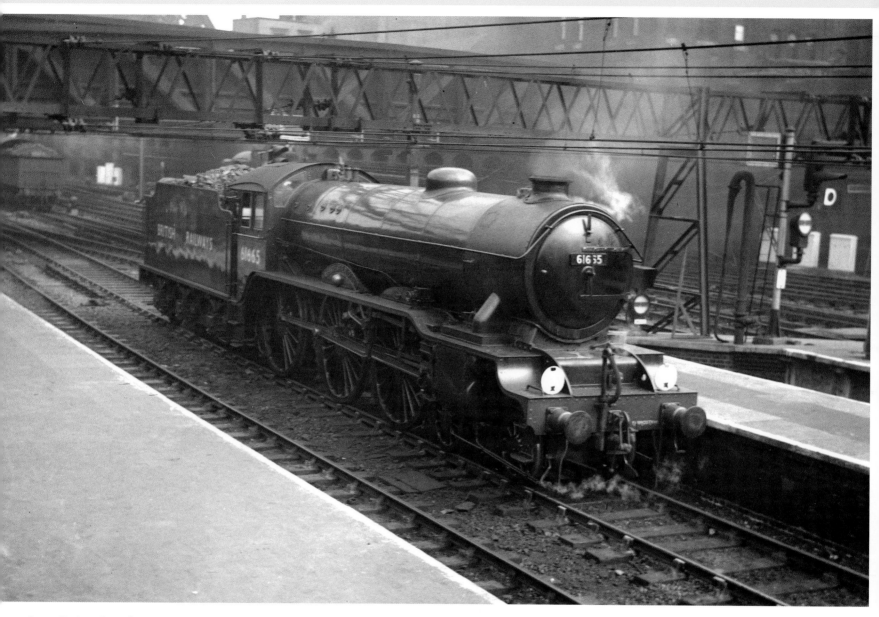

Opposite top: Saturday 8 May 1948. At Stratford shed, ex-LNER three-cylinder Class B17 4-6-0 No. 1634 *Hinchingbrooke* is parked in the yard. Constructed during 1931 at Darlington Works, she would be withdrawn from service in 1958.

Opposite bottom: Saturday 8 May 1948. At Stratford station, Class L1 2-6-4 tank No. 67708 is working a local suburban train. Constructed at Darlington Works in 1948, she would be withdrawn during 1960.

Above: Saturday 10 July 1948. At Liverpool Street station, ex-LNER three-cylinder class B17 4-6-0 No. 61665 *Leicester City* is waiting to reverse out of the station platform. Constructed by Robert Stephenson & Co. during 1937, she would be withdrawn from service in 1958.

Above: August 1948. This photograph of Ministry of Supply 'Austerity' Class 8F 2-8-0 No. 63127, seen at Neasden shed, is included here as she is bearing an LNER number. Constructed by the Vulcan Foundry in 1944, she spent time in Europe after D Day and would be returned to the UK in 1946, being loaned to the LNER and finally purchased by them during 1947. Originally numbered 78627 by the WD, she would become No. 90448 with British Railways and be withdrawn from service during 1965.

Opposite top: Friday 17 August 1948. Standing in Gorton shed yard awaiting entry to the works is ex-GCR Class 1A (LNER Class B8) 4-6-0 No. 1355. Constructed at the same works in 1914, she would be withdrawn from service during the month following this photograph, September 1948.

Opposite bottom: Friday 17 August 1948. Originally designed with domeless boilers, the ex-H&BR Class G3 (LNER Class J75) 0-6-0 tanks were constructed in two batches, with the first six examples coming from the Yorkshire Engine Co. in 1901 and 1902 and the remaining ten locomotives being the product of Kitson & Co. during 1908. The photographer has again travelled west to this former CLC shed at Walton on the Hill near Liverpool to photograph No. 8365, a Kitson & Co. example. She would be rebuilt with a domed boiler in 1927 and be the last of the class to be withdrawn during January 1949.

Opposite top: Saturday 28 August 1948. At Darlington shed, ex-LNER Class V2 2-6-2 No. 60920 has recently emerged from the works after an overhaul and painting in British Railways livery with its new number. Constructed at the same works in 1941, she would be withdrawn during 1962.

Opposite bottom: Saturday 18 September 1948. Designed by John Robinson to handle the suburban traffic out of Marylebone station, his Class 9N 4-6-2 tanks for the GCR saw twenty-one examples exiting Gorton Works between 1911 and 1917. A further ten examples were constructed at Gorton Works during 1924 and 1925 with Hawthorn, Leslie & Co. supplying thirteen more examples during 1925 and 1926. Designated Class A5 by the LNER, the massive bulk of No. 9801 is seen here at Neasden shed. The second example to enter service in 1911, she would be withdrawn during 1960.

Above: Saturday 18 September 1948. Ex-GCR Class 9F (LNER Class N5) 0-6-2 tank No. 9369 is seen manoeuvring in Neasden shed yard. Entering service from Beyer Peacock & Co. as the penultimate member of the class during 1901, she would be withdrawn from service in 1957.

Opposite top: Saturday 18 September 1948. Also seen in Neasden shed yard is classmate N5 No. E9318, which was a product of Gorton Works from 1899 that would be withdrawn during 1956.

Opposite bottom: Saturday 15 January 1949. At Stratford Works, ex-LNER Class B2 4-6-0 No. 61603 *Framlingham* is looking sparkling in her new British Railways livery. Originally constructed by the NBL in 1928 as a three-cylinder Class B17 locomotive, she would be rebuilt at Darlington Works during the Edward Thompson period in 1946 as a two-cylinder locomotive and classified Class B2. A total of only ten examples of the Class B17 were so rebuilt between 1945 and 1949.

Above: Saturday 7 May 1949. In Stratford shed yard ex-GER Class C53 (LNER Class J70) six-wheeled tram engine No. 8217 is awaiting its new owner's identity and number. Constructed at Stratford Works during 1903, she would be numbered 136 by the GER and sent to work in the Ipswich Dock area. Becoming No. 7136 and later No. 8217 with the LNER, she would be numbered 68217 with British Railways and be withdrawn from service in 1953.

Opposite top: Saturday 23 July 1949. Seen at Stratford Works is one of only five class members constructed, ex-GER Class B74 (LNER Class Y4) 0-4-0 tank No. 68129 had entered service from the same works in 1921 and spent its entire life as one of the works shunters. Numbered 210 by the GER and later No. 7210 with the LNER, she also carried the Departmental No. 33 before becoming No. 68129 with British Railways. She would be the last of the class to be withdrawn from service in 1963.

Opposite bottom: Thursday 11 August 1949. This diminutive locomotive, seen here at York shed yard, is ex-NER Class K (LNER Class Y8) 0-4-0 tank No. 8091. Constructed at Gateshead Works during 1890, she would be transferred to Departmental Stock in 1954 and numbered 55. She would be withdrawn after sixty-six years of service during 1956.

Above: Saturday 24 September 1949. Another Class J70 tram locomotive seen here in Stratford shed yard is No. 68223, sporting its new owner's identity. Constructed at Stratford Works during 1914, she would be numbered 131 by the GER becoming No. 7131 and later No. 8223 with the LNER. She would be withdrawn during 1955.

Saturday 3 June 1950. Approaching York station at the head of the 'up', 'The Northumbrian' is as yet unnamed Class A1 4-6-2 No. 60157 sporting the blue livery with which she entered service from Doncaster Works seven months earlier. Named *Great Eastern* in 1951, she was also repainted in the British Railways green livery at the same time. She would only give sixteen years of service, being withdrawn during 1965. 'The Northumbrian' service between Newcastle and London commenced during 1949 and would cease in 1964.

Saturday 3 June 1950. Working as one of the York station pilots on this day is Class J72 0-6-0 tank No. 69020. One of the post-nationalisation examples of the class constructed at Darlington Works and entering service five months earlier, she would be withdrawn after only thirteen years of work in 1963.

This class entered service with the NER during 1898 as their Class E1, with construction continuing until 1951 when the final eight members of the class entered service from Darlington Works.

Opposite top: Saturday 3 June 1950. Approaching York station at the head of an 'up' express is Class B1 4-6-0 No. 61337. Entering service from the NBL in Glasgow during 1948, she would be withdrawn only nineteen years later in 1967.

Opposite bottom: Saturday 3 June 1950. Another York station pilot seen on this day is ex-NER Class E (LNER Class J71) 0-6-0 tank No. 68286. Compare this photograph with that of the same locomotive on page 9 sporting full LNER lined livery.

Above: Sunday 4 June 1950. This 'up' express departing from York station is seen accelerating past the former Racecourse platform behind ex-LNER Class A2/3 4-6-2 No. 60511 *Airborne*. This Edward Thompson 'Standard' Pacific design for the LNER saw only fifteen examples constructed, all at Doncaster Works during 1946 and 1947. No. 60511 entered service during 1946 and was named after the racehorse that won the 1946 Derby and St Leger races. She would spend the bulk of her working life allocated to Heaton shed in Newcastle and be withdrawn during 1962.

Sunday 4 June 1950. In sparkling condition and wearing the experimental blue livery applied the previous year, ex-LNER Class A3 4-6-2 No. 60072 *Sunstar* is seen passing the former Racecourse platform at York at the head of an 'up' express. Constructed by the NBL during 1924 as a Class A1 locomotive, she would be rebuilt as a Class A3 in 1941. Named after the racehorse that won the 1911 Derby and 2000 Guineas races, much of her working life was based at Heaton shed in Newcastle. She would be withdrawn from service in 1962. Note the wonderful signal gantry bearing both lower and upper quadrant signals that would soon disappear.

Saturday 10 June 1950. Seen near Potters Bar at the head of the 'up' summer-only service 'The Scarborough Flyer' is Class A2 4-6-2 No. 60526 *Sugar Palm*. Entering service from Doncaster Works during 1948, she would acquire the double chimney seen here in 1949. Allocated to York shed for the bulk of her working life and named after the racehorse that won the 1944 Nunthorpe Stakes, she would be withdrawn during 1962.

Saturday 10 June 1950. In terrible external condition, ex-LNER Class A3 4-6-2 No. 60089 *Felstead* is working a 'down' express near Potters Bar. Named after the racehorse that won the 1928 Derby, she was constructed at Doncaster Works during 1928 and would, after an overhaul in October 1950, sport the British Railways blue livery, which she would lose during 1952 in reverting to the green livery. Acquiring a double chimney in 1959 and trough-style smoke deflectors during 1961, she would be withdrawn from service in 1963.

Saturday 10 June 1950. Seen near Potters Bar at the head of a 'down' express is the final member of the Arthur Peppercorn-designed Class A2 4-6-2 No. 60539 *Bronzino.* Entering service during 1948 from Doncaster Works, she would only give fourteen years of work and be withdrawn from service in 1962.

Saturday 10 June 1950. Near Potters Bar, Class A1 4-6-2 No. 60133 *Pommern* is in charge of the 'down' 'The West Riding' London to Bradford summer-only service. Constructed at Darlington Works during 1948 and named after the racehorse that won the 1915 Derby, 2000 Guineas and St Leger races, she is seen in her British Railways blue livery. Allocated to Copley Hill shed in Leeds at this time, she would be withdrawn during 1965.

Saturday 10 June 1950. Sporting the 'Yorkshire Pullman' headboard and with the fireman attempting to lay a dense smokescreen, ex-LNER Class A4 4-6-2 No. 60032 *Gannet* is working hard with the 'down' train again near Potters Bar. Entering service in 1938 from Doncaster Works, she is seen here in the British Railway blue livery. She would be withdrawn from service in 1963.

Opposite top: Thursday 27 July 1950. In Doncaster Works yard, a rather forlorn-looking ex-GNR Class C1 (LNER Class C1) 4-4-2 No. 2817 had been withdrawn from traffic a month earlier. A product of the same works during 1904, this class of locomotive was the first utilising the 'Atlantic' wheelbase in the UK.

Opposite bottom: Friday 28 July 1950. Bearing a 36C Frodingham shed code, ex-GCR Class 8K (LNER Class O4) 2-8-0 No. 63617 is standing over the ash pits at Gorton shed. Constructed at Gorton Works in 1912, she would be withdrawn during 1962.

Above: Saturday 29 July 1950. Standing in Darlington shed yard and sporting the new British Railways logotype on her tender, ex-NER Class P1 (LNER Class J25) 0-6-0 No. 65691 is looking in good clean condition. Entering service from Gateshead Works during 1900 and fitted with a saturated steam boiler, she would be rebuilt with a superheated boiler in 1916 and be withdrawn during 1961.

Opposite top: Saturday 31 March 1951. In ex-works condition standing in Doncaster shed yard is ex-LNER Class K3 2-6-0 No. 61945. Constructed by the NBL in 1935, she would be withdrawn from service during 1967.

Opposite bottom: Saturday 31 March 1951. Also seen in Doncaster shed yard in ex-works condition is ex-GNR Class C2 (LNER Class C12) 4-4-2 tank No. 67353. Entering service from Doncaster Works during 1898 and numbered 1014 by the GNR, she would become No. 4014 and later No. 7353 with the LNER. She would be withdrawn from service in 1955.

Above: Saturday 31 March 1951. Bearing a 36B Mexborough shed code and standing at the head of an 'up' express at Doncaster station is ex-LNER Class B1 4-6-0 No. 61168. An example of the Vulcan Foundry-constructed members of the class from 1947, she would only give eighteen years of service, being withdrawn during 1965.

Saturday 14 April 1951. This day saw the running of the 'East London No. 1 Rail Tour', organised by the RCTS London Branch. Departing from Fenchurch Street station, it visited Bromley, North Woolwich and Stepney, ending at Stratford. Seen standing beside North Woolwich 'box is motive power for part of the day, ex-GER Class S56 (LNER Class J69) 0-6-0 tank No. E8619 in green livery, looking splendidly cleaned by the staff at Stratford shed. Constructed at Stratford Works during 1904, she would be withdrawn from service in 1961.

May 1951. Near Potters Bar, ex-LNER Class V2 2-6-2 No. 60880, seen working a north-bound suburban train, is seriously in need of a good clean. Bearing a 36A Doncaster shed code, she was constructed at the works there in 1940. Being fitted with a double chimney in 1961, she would be withdrawn during 1963.

Opposite top: May 1951. Seen near Potters Bar at the head of another north-bound suburban working consisting of two, quad articulated sets is ex-GNR Class N2 (LNER Class N2) 0-6-2 tank No. 69529. A product of the NBL during 1921, she would be withdrawn from service in 1962.

Opposite bottom: Saturday 16 June 1951. Having entered service from the NBL in Glasgow during the month prior to this photograph, Class B1 4-6-0 No. 61377 is seen at Nottingham Victoria station. Bearing a 38A Colwick shed code, she would end her short working life of only eleven years whilst based at Sheffield Darnall in 1962.

Above: Saturday 16 June 1951. This day saw the running of the 'Nottingham Rail Tour', organised by the RCTS East Midlands Branch, behind ex-GNR Class C2 (LNER Class C12) 4-4-2 tank No. 67363. Seen here running round its train at Bestwood, she is bearing a 38B Annesley shed code. Entering service from Doncaster Works in 1899, she would be withdrawn from service during 1958.

Opposite top: Friday 27 July 1951. Bearing a 40A Lincoln shed code, ex-LNER Class J39 0-6-0 No. 64887 is seen standing in its home shed yard. Constructed at Darlington Works during 1935, she would be withdrawn from service in 1960.

Opposite bottom: Saturday 28 July 1951. At Hull Dairycoates shed, ex-GCR Class 8K (LNER Class O4) 2-8-0 No. 63603 is carrying the No. 16 pilot board. Constructed at the former GCR Gorton Works in 1913, she would be withdrawn from service after forty-nine years during 1962.

Above: Saturday 28 July 1951. Class L1 2-6-4 tank No. 67763 is seen at its home shed, 53B Hull Botanic Gardens. Constructed by the NBL in 1949, she would have a short working life of only thirteen years, being withdrawn during 1962. Ninety-nine examples of the class were constructed between 1948 and 1950, with thirty-five coming from the NBL, thirty-five from Robert Stephenson & Hawthorns and twenty-nine from Darlington Works. These, together with the initial No. 9000 from Doncaster Works in 1945, gives a total of 100 locomotives in the class.

Saturday 28 July 1951. Seen standing in Hull Botanic Gardens shed yard is ex-NER Class W (LNER Class A6) 4-6-2 tank No. 69791. She was originally constructed at Gateshead Works as an example of a class comprising only ten locomotives with a 4-6-0 wheel arrangement to handle the traffic on the route between Scarborough and Whitby. All were rebuilt as 4-6-2 tanks between 1914 and 1917 and given larger coal bunkers. No. 69791 entered service during 1907, would be rebuilt in 1917 and withdrawn from service during the month following this photograph, August 1951.

Saturday 28 July 1951. The Nigel Gresley-designed three-cylinder 4-4-0s of Class D49 appeared utilising three differing types of valve operation. Part 1 used Walschaerts gear, Part 2 was fitted with Lentz rotary cam-operated poppet valves and Part 3 utilised Lentz oscillating cam-operated poppet valves. Seen at Hull Botanic Gardens shed is a member of the Part 1 group, No. 62703 *Hertfordshire*. Entering service during 1927 from Darlington Works, she would be withdrawn in 1958.

Wednesday 1 August 1951. An example of only twelve locomotives constructed in the class, ex-GCR Class 9L (LNER Class C14) 4-4-2 tank No. 67441 is seen at Doncaster shed. Constructed in 1917 by Beyer, Peacock & Co. with a saturated steam boiler, she would be rebuilt during 1926 with a superheating boiler and be withdrawn from service in 1957.

Friday 10 August 1951. At Gateshead shed, ex-NER Class E (LNER Class J71) 0-6-0 tank No. 8251 has still to receive its new owner's identity and number. Constructed during 1889 at Darlington Works and numbered 501 by the NER, she would become No. 8251 with the LNER and would give seventy years of service, being withdrawn during 1959.

Saturday 11 August 1951. This spread shows three examples of the ex-NER Class P2 (LNER Class J26) 0-6-0s, designed by Wilson Worsdell and introduced during 1904. All are seen at Newport shed.

Opposite top: No. 65774 was constructed at Gateshead Works in 1905 and numbered 525 by the NER, becoming No. 5774 with the LNER. Fitted with a smaller chimney and round spectacle glasses, she would be withdrawn during 1961.

Opposite bottom: Another Gateshead Works-constructed member of the class, No. 65752 entered service in 1905, numbered 816 by the NER and later No. 5752 by the LNER. She still retains the taller chimney and oblong spectacle glasses and would be withdrawn during 1958.

Above: No. 65732 was a Darlington Works-constructed example from 1904 that would be withdrawn during 1959. She also still retains the taller chimney and oblong spectacle glasses.

Saturday 2 February 1952. This spread of three photographs taken at the photographer's local station, New Southgate, shows a sequence of 'up' workings to King's Cross.

Above: Ex-LNER Class A3 4-6-2 No. 60053 *Sansovino* entered service from Doncaster Works during 1924 as a Class A1 locomotive. She would be rebuilt as a Class A3 locomotive in 1943 and fitted with a double chimney during 1958. Named after the racehorse that won the 1924 Derby, she would be withdrawn from service in 1963.

Opposite top: Ex-LNER Class V2 2-6-2 No. 60920 had been constructed at Darlington Works in 1941 and would be withdrawn from service during 1962.

Opposite bottom: Ex-LNER Class A4 4-6-2 No. 60015 *Quicksilver* was the second member of the class to enter service from Doncaster Works in 1935. Numbered 2510 by the LNER and later becoming No. 15, she would end her days based at King's Cross shed and be withdrawn during 1963.

Saturday 2 February 1952. A further three photographs are seen here taken at New Southgate station. All show 'up' passenger workings.

Opposite top: At the head of the 'up', 'The West Riding' Class A1 4-6-2 No. 60141 *Abbotsford* is seen powering through the station. Entering service from Darlington Works in 1948, she would only give sixteen years of service, being withdrawn during 1964. Abbotsford is the name of the house built by Sir Walter Scott near Melrose.

Opposite bottom: Class A1 4-6-2 No. 60121 *Silurian* had been constructed at Doncaster Works in 1948 and was named after the racehorse that had won the 1923 Doncaster Cup. She would be withdrawn during 1965.

Above: Ex-LNER Class V2 2-6-2 No. 60842 was the product of Darlington Works in 1938 and would be withdrawn during 1962.

Opposite top: Saturday 2 February 1952. Also seen at New Southgate station on the same day on an 'up' Newcastle United football special is ex-LNER Class A2/3 4-6-2 No. 60515 *Sun Stream*. Named after the racehorse that won the 1945 Oaks and 1000 Guineas, she had been the product of Doncaster Works during 1946 and would be withdrawn from service in 1962. Newcastle United were playing away on this day against Tottenham Hotspur.

Opposite bottom: Sunday 11 May 1952. This day saw the running of the 'South Yorkshire No. 1 Rail Tour', organised by the RCTS Sheffield Branch. Motive power used was ex-LNER Class B1 4-6-0 No. 61166, which is seen here waiting to depart Shireoaks station on its return to Sheffield. A product of the Vulcan Foundry during 1947, she would be withdrawn from service in 1966.

Above: Thursday 5 June 1952. Standing in a row of locomotives at Stratford shed is ex-GER Class R24 (LNER Class J69) 0-6-0 tank No. 8568. Constructed at Stratford Works during 1896 and numbered 392 by the GER, she would become No. 7392 and later No. 8568 with the LNER. Numbered 68568 with British Railways, she would be withdrawn after sixty-two years' service in 1958.

Saturday 7 June 1952. Designed to handle the GER suburban traffic around London, their Class M15 (LNER Class F5) 2-4-2 tanks were all constructed at Stratford Works between 1884 and 1909. Seen here at Stratford shed is No. 67202, which had entered service in 1905 and would be fitted with push-pull gear in 1949 to enable working on the Epping to Ongar branch. Withdrawal would come in 1957.

Saturday 5 July 1952. In Norwich shed yard, ex-GER Class D56 (LNER Class D16) 4-4-0 No. 62556 is parked in a row of locomotives. Constructed at Stratford Works during 1906 she would be rebuilt with a superheating boiler in 1914 and further rebuilt during 1929 in the form seen here. Numbered 1845 by the GER, she would become No. 8845 and later No. 2556 with the LNER and be withdrawn during 1957.

Saturday 5 July 1952. Ex-GER Class T77 (LNER Class J19) 0-6-0 No. 64673 is seen at Norwich shed. These powerful, superheated boiler locomotives were introduced during 1912, with the example seen here entering service from Stratford Works as the penultimate member of the class in 1920. She would be withdrawn during 1962.

Saturday 5 July 1952. In Norwich shed yard stands ex-GER Class D56 (LNER Class D16) 4-4-0 No. 62576. A product of Stratford Works in 1909 incorporating a saturated boiler, she would be rebuilt by the GER during 1922 with a superheating boiler. Rebuilt once more by the LNER in 1937 and becoming a Class D16 locomotive, she would be withdrawn from service in 1957.

Opposite: Saturday 5 July 1952. At Norwich Thorpe station, ex-LNER Class N7 0-6-2 tank No. 69708 is waiting to depart with a working to Yarmouth. Constructed post-grouping at Doncaster Works in 1927, the design was based on the earlier Great Eastern Railway Class L77 locomotives introduced in 1915. She would be withdrawn from service in 1961.

Above: Friday 18 July 1952. Waiting to depart with the two-coach Framlingham branch train at Parham station is ex-GER Class G69 (LNER Class F6) 2-4-2 tank No. 67230, carrying a 32B Ipswich shed code. Entering service from Stratford Works during 1911, she would be withdrawn in 1958. The 6½-mile-long Framlingham branch from its junction at Wickham Market saw only four trains daily each way and services would be withdrawn from service four months later in November 1952.

Saturday 19 July 1952. Standing in the bay platform at Beccles station is ex-GER Class Y14 (LNER Class J15) 0-6-0 No. 65471. A member of this most numerous class of locomotives, a total of 272 were constructed by the GER between 1883 and 1913. No. 65471 entered service in 1913 from Stratford Works. She was numbered 543 by the GER, becoming No. 7543 and later No. 5471 with the LNER. She would be withdrawn from service in 1960.

Saturday 19 July 1952. At Haughley station ex-GER Class Y14 (LNER Class J15) 0-6-0 No. 65467 waits in the bay platform at the head of the Mid-Suffolk Laxfield branch train. Entering service during 1912 from Stratford Works, she was numbered 569 by the GER, becoming No. 7569 and later No. 5467 with the LNER. She would be withdrawn from service in 1959. The Mid-Suffolk branch would close at the end of July 1952, with the station at Brockford and Wetheringsett becoming the base for the heritage Mid-Suffolk Light Railway.

Saturday 19 July 1952. What the photographer describes in his notes as the Laxfield branch locomotive is seen here in the shed at Laxfield. Ex-GER Class Y14 (LNER Class J15) 0-6-0 No. 65447 was constructed at Stratford Works in 1899 and would be withdrawn after sixty years of service during 1959. The shed looks on the point of collapse. It soon would, as the branch closed at the end of July 1952.

Opposite: August 1952. Originally constructed at Darlington Works by the NER in 1921 as a three-cylinder Class H1 4-4-4 tank, No. 69890, seen here at Whitby shed, would be rebuilt during 1935 as a 4-6-2 tank and designated Class A8 by the LNER. The 4-4-4s were found to be lacking in adhesion and their rebuilding as 4-6-2s led them to be more sure-footed. No. 69890 would be withdrawn from service during 1958.

Above: August 1952. Another Class A8 4-6-2 seen here at Whitby shed is No. 69861, which had been constructed at Darlington Works in 1913 as a Class H1 4-4-4 tank and rebuilt as a 4-6-2 tank in 1935. She was destined to be withdrawn during 1960.

Opposite top: Wednesday 20 August 1952. In Boston shed yard is an example of the ex-GNR Class J21 (LNER Class J2) 0-6-0, of which only ten examples were constructed at Doncaster Works during 1912. Designed by Henry Ivatt just prior to Nigel Gresley becoming CME, the class was destined to handle fast goods traffic, with several finding their way to Boston and Lincoln sheds. No. 65020 would be the last of the class to be withdrawn from service in 1954.

Opposite bottom: Wednesday 20 August 1952. The GNR Class O2 (LNER Class O2) 2-8-0 was the first of Nigel Gresley's designs for a locomotive incorporating three cylinders. The first example appeared during 1918, but it was 1921 before further locomotives of the class were constructed, with a total of sixty-seven entering service, the last in 1943. No. 63937 is seen here at Frodingham shed. A Doncaster-constructed example from 1923, she would be withdrawn during 1963.

Above: Wednesday 20 August 1952. Also in Frodingham shed yard is ex-GCR Class 8K (LNER Class O4) 2-8-0 No. 63745. Constructed by the NBL during 1912, she would be rebuilt in 1932 with an improved boiler and designated Class O4/5 by the LNER. Numbered 1207 by the GCR and later No. 3554 and No. 3745 by the LNER, she would be withdrawn from service during 1959.

Saturday 6 September 1952. The branch line to Thaxted was opened in 1913 and closed just short of forty years later in 1952. The two photographs here, taken a week before closure of passenger services, show ex-GER class R24 (LNER Class J69) 0-6-0 tank No. 68579 taking water whilst standing outside the shed at Thaxted **(opposite top)**, and No. 68579 waiting to depart with a train to the junction at Elsenham **(opposite bottom)**. Constructed at Stratford Works during 1896, No. 68579 would be withdrawn from service in 1960. The two coaches used as the branch train were originally constructed as ambulance vehicles by the GER for use during the First World War. These were converted to passenger coaches at the end of the hostilities.

Above: Saturday 10 January 1953. Purchased by the LNER from the Sentinel Co. during 1929, Class Y1 No. 68148 was allocated to Bridlington shed and would remain based there for her entire working life, being withdrawn from service in 1955. She is seen here at Bridlington shed parked adjacent to a 16-ton mineral wagon, with a crew member loading coal into her bunkers.

Opposite top: Saturday 14 March 1953. At Neasden shed ex-LNER Class A3 4-6-2 No. 60111 *Enterprise* is seen sporting green livery applied earlier in the same month. Constructed at Doncaster Works during 1923 as a Class A1 locomotive, she would be rebuilt as a Class A3 in 1927, acquiring a double chimney in 1957 and trough-style smoke deflectors during 1962 – only to be withdrawn later in that year.

Opposite bottom: Saturday 14 March 1953. The nameplate of No. 60111 *Enterprise* was photographed on the same day. She was named after the racehorse that won the 2000 Guineas race in 1887.

Above: Saturday 21 March 1953. Seen here again is ex-GER Class S56 (LNER Class J69) 0-6-0 No. E8619 at Stratford shed, looking in a much dirtier condition to that in the photograph on page 46.

Saturday 21 March 1953. At Stratford shed, ex-GER Class Y14 (LNER Class J15) 0-6-0 No. 65420 is carrying a 31E Bury St Edmunds shed code. Constructed at Stratford Works during 1892, she would give seventy years of service, being withdrawn in 1962.

Friday 10 April 1953. At Hull Botanic Gardens shed, ex-NER Class U (LNER Class N10) 0-6-2 tank No. 69093 is seen bearing a 53A Hull Dairycoates shed code. A product of Darlington Works in 1902, she would be withdrawn during 1957.

Friday 10 April 1953. Bearing a 53A Hull Dairycoates shed code, ex-LNER Class K3 2-6-0 No. 61819 is waiting to depart from Hull Paragon station at the head of an 'up' express. Constructed at Darlington Works during 1924, she would be withdrawn from service in 1962.

Sunday 7 June 1953. This day saw the running of the RCTS Sheffield Branch 'South Yorkshire No. 2 Rail Tour', which departed from Sheffield Midland and visited Shireoaks, Mexborough, Elsecar and Penistone, returning to Sheffield Midland. One of the locomotives working that day was ex-GCR Class 11F (LNER Class D11) 4-4-0 No. 62667 *Somme* and she is seen here.

At Elsecar yard **(opposite top)** and a close-up of her nameplate **(opposite bottom)**. Constructed at Gorton Works in 1922, she would be withdrawn from service during 1960.

Above: The Elsecar Junction onto the Elsecar branch was handled by ex-GCR Class 9J (LNER Class J11) 0-6-0 No. 64374, which is seen here at the sidings in Elsecar. A product of the Vulcan Foundry in 1904, she would be withdrawn from traffic during 1955.

Above: Monday 3 August 1953. Designed by Nigel Gresley specifically to handle the heavy passenger traffic on the Great Eastern section of the LNER, his three-cylinder Class B17 4-6-0s were introduced in 1928. A total of seventy-three examples were constructed by the NBL in Glasgow, Robert Stephenson & Co. and Darlington Works, with the last example entering service during 1937. No. 61657 *Doncaster Rovers* is seen here at Hitchin station working an 'up' Cambridge Buffet express. A Darlington Works example from 1936, she would be withdrawn from service in 1960.

Opposite top: Saturday 8 August 1953. Approaching Potters Bar station at the head of the 'down', 'The West Riding' comprising of a rake of coaches containing some articulated stock, is ex-LNER Class A3 4-6-2 No. 60058 *Blair Athol*. Constructed at Doncaster Works during 1925 as a Class A1 locomotive and named after the racehorse that won the 1864 Derby and St Leger races, she would be rebuilt in 1945 as a Class A3 locomotive and be withdrawn from service in 1963.

Opposite bottom: Sunday 8 August 1953. Seen near Potters Bar, with what the destination board tells us is a 'down' working to Hatfield, ex-LNER Class N2 0-6-2 tank No. 69579 has the usual pair of Quad Articulated sets in tow. A product of Hawthorn, Leslie & Co. during 1929, she would be withdrawn from service in 1962. (AGF366B)

Opposite top: Tuesday 25 August 1953. At Selby shed, ex-NER Class T2 (LNER Class Q6) 0-8-0 No. 63449 is simmering gently. Constructed by Armstrong Whitworth & Co. in 1920, she would be withdrawn during 1963.

Opposite bottom: Saturday 29 August 1953. Standing in Darlington shed yard in ex-works condition is ex-NER Class P3 (LNER Class J27) 0-6-0 No. 65838. Constructed by Robert Stephenson & Co. in 1909, she would be withdrawn from service in 1967.

Above: Friday 4 September 1953. Designed by Wilson Worsdell as an express passenger class for the NER and introduced during 1899, their Class R 4-4-0s consisted of sixty examples, all constructed at Gateshead Works, with the last appearing in 1907. Designated Class D20 with the LNER, No. 62372 is seen here in Tyne Dock shed yard. She had entered service in 1906 and would be withdrawn fifty years later during 1956.

Opposite: Sunday 20 September 1953. To commemorate the opening of Doncaster Works by the Great Northern Railway in 1853, a special train was organised to run from London King's Cross to Doncaster on this day, to be named *The Plant Centenarian*. Motive power would be the two preserved ex-GNR 'Atlantics' Nos 251 and 990 *Henry Oakley*. No. 990 had the distinction of being the very first 'Atlantic' wheelbase locomotive to be constructed in Great Britain, entering service from Doncaster Works in 1898. She was withdrawn from traffic during 1937 and placed in the York Museum in 1938. Seen here at King's Cross prior to departure, No. 990 is coupled in front of No. 251.

Above: Sunday 20 September 1953. This photograph shows the right-hand driving wheels and nameplate on ex-GNR Class C1 (LNER Class C2) 4-4-2 No. 990 *Henry Oakley*, named after the General Manager of the GNR, who served in that post from 1870 until 1898. The locomotive was named in his honour during 1900.

Above: Saturday 3 October 1953. In Hornsey shed yard, the driver is seen filling the cylinder oil pots on ex-LNER Class J50 0-6-0 No. 68945. Constructed at Doncaster Works during 1926, she would be withdrawn from service in 1961. This example was a post-grouping member of the class of seventy-two locomotives that were introduced as GNR Class J23 by Nigel Gresley in 1922, with construction continuing until 1939.

Opposite top: Saturday 3 October 1953. Standing in Hornsey shed yard is ex-GNR Class N2 (LNER Class N2) 0-6-2 tank No. 69505. Constructed by the NBL during 1920 and numbered 1726 by the GNR, she would become No. 4726 and later No. 9505 with the LNER. She would give forty years of service, being withdrawn in 1960.

Opposite bottom: Saturday 3 October 1953. Also seen in Hornsey shed yard is an example of the earlier class of GNR 0-6-2 tank, GNR Class N1 (LNER Class N1). No. 69435 had been constructed at Doncaster Works during 1907 with a saturated steam boiler and incorporating condensing gear. Rebuilt in 1924 with a superheating boiler, the condensing gear was retained as seen here. Numbered 1555 by the GNR, she would become No. 4555 and later No. 9435 with the LNER. She would be withdrawn during 1955.

Opposite top: 10 April 1954. At Retford shed, ex-GNR Class J4 (LNER Class J3) 0-6-0 No. 64140 is coaled up and awaiting its next duty. Constructed during 1900 at Doncaster Works, she would be withdrawn from service eight months after this photograph, in December 1954. This numerous class was originally introduced during 1873 to a design by Patrick Stirling. It was found so useful that a total of 240 examples were constructed, the last appearing in 1901. One-hundred and fifteen examples entered service from Doncaster Works, with ninety coming from Dübs & Co., twenty-four from Kitson & Co. and eleven from the Vulcan Foundry.

Opposite bottom: Saturday 10 April 1954. At the former GNR shed at Retford, ex-GCR Class 9C (LNER Class N5) 0-6-2 tank No. 69277 has just completed taking water. Constructed by Beyer, Peacock & Co. during 1894, she would be numbered 544 by the GCR, becoming No. 5544 and later No. 9277 with the LNER. She would give sixty-two years of service, being withdrawn in 1956.

Above: Monday 19 April 1954. Seen entering Oakleigh Park station with an 'up' working to King's Cross is Class L1 2-6-4 tank No. 67793. A product of Robert Stephenson & Co. in 1950, she would have a short working life of only twelve years, being withdrawn during 1962.

Above: Saturday 10 July 1954. In contrast to the photograph on page 74 taken just over a year earlier, Class A3 No. 60111 *Enterprise*, bearing a 34E New England shed code, is looking uncared for. Seen at Marylebone station, she is waiting for parcels to be loaded and passengers to board before departing at the head of a north-bound express working.

Opposite top: Saturday 21 August 1954. Ex-GER Class B74 (LNER Class Y4) 0-4-0 tank No. 68128 is seen here at Stratford shed. Constructed at Stratford Works during 1921, this powerful short wheelbase tank locomotive found use in very restricted yard situations and would be withdrawn from service in 1956.

Opposite bottom: Friday 3 September 1954. Seen at Darlington shed, ex-NER Class C (LNER Class J21) 0-6-0 No. 65064 had been constructed during 1890 at Gateshead Works, utilising a saturated steam boiler. Being rebuilt in 1915 with a superheating boiler, she would be withdrawn from service in 1958.

Friday 3 September 1954. Ex-NER Class S3 (LNER Class B16) 4-6-0 No. 61466 is in ex-works condition standing in Darlington shed yard. Designed by Vincent Raven and introduced during 1919 for the NER as a mixed traffic class, these three-cylinder locomotives were so useful that further examples were constructed post-grouping. No. 61466 was a Darlington Works example from 1923 that would be withdrawn during 1961.

Sunday 5 September 1954. Seen at Doncaster station is the final example of the Arthur Peppercorn design of Class A1 4-6-2, No. 60162 *Saint Johnstoun*. Entering service from Doncaster Works in December 1949 and allocated new to Haymarket shed in Edinburgh, she spent her entire working life based there and would be withdrawn from service in 1963. The name is based on the alternative locally used name for the city of Perth. None of the forty-nine examples of the class made it to the preservation scene, but the A1 Steam Locomotive Trust successfully constructed the fiftieth member in 2008, No. 60163 *Tornado*, which is seen regularly working main line specials.

Above: Saturday 5 March 1955. Standing beside a water crane in Cambridge shed yard is ex-GER Class T26 (LNER Class E4) 2-4-0 No. 62784. Constructed at Stratford Works during 1894 and numbered 478 by the GER, she would become No. 7478 and later No. 2784 with the LNER. Note that she is coupled to a larger-capacity 'water-cart' style of tender, which would remain with her until withdrawal two months later, in May 1955.

Sunday 20 March 1955. **Opposite:** At Sheffield Darnall shed, a comparison is here seen of two locomotives of the former GCR Class 8K (LNER Class O4) 2-8-0.

Top: No. 63604 had originally been constructed at Gorton Works in 1913 and would be rebuilt under the Edward Thompson regime, with a new round-top boiler and cab, during 1954.

Bottom: No. 63852 is still in as-built condition with a Belpaire firebox. Entering service from the NBL in Glasgow in 1919 under an order from the Ministry of Munitions, she was numbered 2054 with the ROD and would be purchased by the LNER in 1924. This example would be rebuilt eight months after this photograph, in November 1955, with a new boiler and cab. Both locomotives would be withdrawn from service during 1964.

Saturday 9 July 1955. On the approach to Potters Bar station, Class A1 4-6-2 No. 60156 *Great Central* is at the head of a 'down' express. Constructed at Doncaster Works during 1949, she would be allocated new to Gateshead shed and would end her days during 1965 whilst based at York shed. The Arthur Peppercorn design of Class A1 locomotive for the LNER did not enter service until after the nationalisation of the railways in 1948. The construction of the class was split between Doncaster and Darlington Works, with a total of forty-nine examples being produced. Doncaster Works constructed twenty-six locomotives with Darlington contributing twenty-three members of the class. They gained a reputation as free steaming locomotives that could handle the heaviest of loads and were allocated throughout the former LNER territory, except for the former GE section.

Saturday 23 July 1955. The Ministry of Supply 'Austerity' 0-6-0 saddle tanks, produced for use with the War Department, started life as an existing Hunslet Engine Co. design dating back to 1937. A total of 377 examples were delivered to the War Department between 1943 and 1947, with many being constructed by other independent suppliers – Andrew Barclay Sons & Co., W.G. Bagnall & Co., Hudswell Clarke & Co., Robert Stephenson & Hawthorns and the Vulcan Foundry all contributing. In 1946 the LNER purchased a total of seventy-five examples, which had been constructed between 1944 and 1946. No. 68017, seen here in Sunderland shed yard, was a product of the Hunslet Engine Co. from 1944 that would be withdrawn during 1962. Construction of the locomotives continued after the war, with further examples being ordered by the National Coal Board and other industrial users, the last entering service in 1964. Many examples have entered the preservation scene, with approximately seventy seeing service with heritage railways.

Above: Saturday 23 July 1955. The Nigel Gresley Class V1 2-6-2 tanks were specifically designed to handle the suburban traffic around Edinburgh, Glasgow and the cities of north-east England. Ninety-two examples were constructed, all at Doncaster Works, between 1930 and 1940. All bar the last ten examples were constructed as Class V1 locomotives with a 180psi boiler pressure. The final ten examples utilised a boiler pressure of 200psi and were designated Class V3. The majority of the Class V1s were rebuilt over a period of years to Class V3s with No. 67652, seen here at Tyne Dock shed, entering service in 1936 and being rebuilt as a Class V3 during 1952. She would be withdrawn from service in 1963.

Opposite top: Monday 8 August 1955. At the head of the 'up', 'The Elizabethan' ex-LNER Class A4 4-6-2 No. 60031 *Golden Plover* is seen passing through Doncaster station. Constructed at Doncaster Works in 1937 and allocated new to Haymarket in Edinburgh, she would acquire a double chimney in 1958 and be transferred to St Rollox in Glasgow to assist with the three-hour workings between Glasgow and Aberdeen. She would be withdrawn from service in 1965.

Opposite bottom: Sunday 21 August 1955. A further 'up' express is seen here entering Oakleigh Park station. At its head is the world speed record holder ex-LNER Class A4 4-6-2 No. 60022 *Mallard.* Entering service from Doncaster Works in March 1938, she acquired the record during a run on 3 July that year. Withdrawn from service during 1963, she can be seen at the National Railway Museum in York.

Sunday 29 April 1956. This day saw the running of the 'SMJR Rail Tour', organised by the SLS, commencing at London King's Cross with a route to Stratford-upon-Avon passing Hitchin, Bedford and Towcester. The return was made via Birmingham New Street, Rugby Midland and Bletchley, arriving at London Euston. Motive power for the day was ex-GER Class D56 (LNER Class D16) 4-4-0 No. 62605, carrying a 31A Cambridge shed code, and seen here heading towards Bowes Park station on the Hertford Loop. Entering service from Stratford Works during 1911, she would be rebuilt on several occasions, becoming a Class D16/3 locomotive in 1940. She would be withdrawn from service in 1957. The Stratford-upon-Avon and Midland Junction Railway (SMJR) connected the Midland and Great Western Railways in Warwickshire with the Great Central and London and North Western Railways in Northamptonshire.

Sunday 27 May 1956. In ex-works condition, ex-GNR Class N2 (LNER Class N2) 0-6-2 tank No. 69529, bearing a 34A King's Cross shed code, is standing in the yard at Doncaster shed. Constructed by the NBL during 1921, she would give forty-one years' service, being withdrawn in 1962.

Sunday 27 May 1956. Designed to handle the increasingly heavy traffic on the Great Eastern section of the LNER, the three-cylinder Class B17 4-6-0s with 6ft 8in driving wheels were introduced during 1928, with a total of seventy-three examples being constructed, the last in 1937. Bearing a 30A Stratford shed code and seen in sparkling ex-works condition at Doncaster shed is the premier example of the class, No. 61600 *Sandringham*. Design and construction of the class were awarded to the NBL in Glasgow during February 1928 and the first ten members of the class entered service in December of that year. *Sandringham* would be withdrawn after thirty years' service in 1958.

Sunday 27 May 1956. Another ex-works locomotive seen standing in Doncaster shed yard is ex-GNR Class J23 (LNER Class J50) 0-6-0 tank No. 68903. Originally constructed at Doncaster Works during 1915 and numbered 171 by the GNR, she would become No. 3171 and later No. 8903 with the LNER. She would be rebuilt in 1932 and be withdrawn during 1961.

Constructed to a design by Nigel Gresley of 1914, the first thirty examples of what became Class J51 with the LNER entered service prior to the grouping. These were all subsequently rebuilt to Class J50 locomotives between 1929 and 1935. Post-grouping, a further seventy-two examples of Class J50 were constructed by both Doncaster and Gorton Works, the last appearing in 1939.

Sunday 27 May 1956. At Doncaster shed, this graceful-looking locomotive, ex-GNR Class C2 (LNER Class C12) 4-4-2 tank No. 67364, has just been withdrawn from service and is awaiting the cutters' torch. Entering service fifty-seven years earlier from Doncaster Works, she had been numbered 1506 by the GNR, becoming No. 4506 and later No. 7364 with the LNER.

Sunday 27 May 1956. In ex-works condition at Doncaster shed is ex-LNER Class B1 4-6-0 No. 61203. Constructed by the NBL during 1947, she would be withdrawn only fifteen years later in 1962.

Sunday 27 May 1956. Also seen in ex-works condition at Doncaster shed is the final member of the Class K1 2-6-0 No. 62070, which entered service from the NBL in 1950. She would only give fifteen years of service, being withdrawn during 1965.

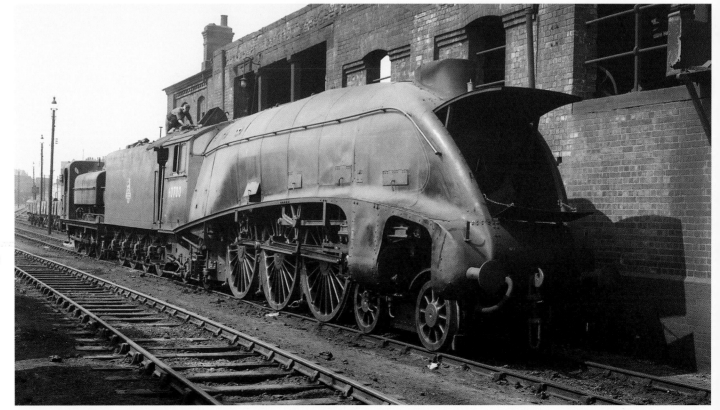

Sunday 27 May 1956. Standing adjacent to the coaling stage at Doncaster shed is the rebuilt ex-LNER Class W1 4-6-4 No. 60700. Originally constructed during 1929 as a four-cylinder compound locomotive with a Yarrow water-tube boiler, she had been rebuilt as a conventional three-cylinder simple with a superheating boiler in 1937. Spending much of her working life allocated to King's Cross shed, she would be withdrawn during 1959.

Above: Sunday 9 September 1956. This day saw the running of the RCTS (London Branch) 'The Fensman No. 2 Rail Tour' commencing at Nottingham and visiting Peterborough, Whittlesea, Spalding, Lincoln and the Upwell Tramway. Motive power on the day included Class B1 4-6-0 No. 61391, which is seen here with its ten-coach train at Whittlesea station. A product of the NBL in 1951, she would be withdrawn after only eleven years during 1962.

Opposite top: Saturday 22 September 1956. Bearing a 34A King's Cross shed code and seen here again is ex-LNER Class W1 4-6-4 No. 60700, whilst working a north-bound express as it enters Wood Green station.

Opposite bottom: Saturday 22 September 1956. Seen about to pass Wood Green station at the head of the 'The West Riding' with this north-bound express is Class A1 4-6-2 No. 60118 *Archibald Sturrock*. Named after the Scots-born engineer who was Locomotive Superintendent of the GNR from 1850 until 1866, this locomotive was constructed at Doncaster Works during 1948. She would spend her entire working life allocated to sheds in Leeds, being withdrawn from service in 1965 whilst based at Neville Hill shed.

Sunday 12 May 1957. This day saw the running of the RCTS (East Midland Branch) 'East Midlander No. 3 Rail Tour', commencing at Nottingham Midland station and visiting Rotherham, Beverley, York, Doncaster and Mansfield returning to Nottingham Midland. Seen here at York station is the motive power on the day, ex-GER Class D56 (LNER Class D16) 4-4-0 No. 62571. Entering service from Stratford Works in 1909, she would be rebuilt on several occasions with improved boilers and be withdrawn from service during 1959.

Saturday 25 May 1957. At Stratford shed, ex-GER class S46 (LNER Class D16) 4-4-0 No. 62526 had been withdrawn from service during the same month. Constructed at Stratford Works during 1902, she would also be rebuilt several times over many years and spend her final working days allocated to March shed 31B, whose shed code she bears.

Opposite top: Sunday 28 July 1957. At Blaydon shed, west of Newcastle, ex-NER three-cylinder Class S3 (LNER Class B16) 4-6-0 No. 61430 is parked in the shed yard. Constructed at Darlington Works during 1921, she would be withdrawn from service in 1959.

Opposite bottom: Sunday 28 July 1957. Also seen in Blaydon shed yard is ex-LNER Class J39 0-6-0 No. 64842. A product of Darlington Works in 1932, she would be withdrawn from service during 1962.

Above: Sunday 7 September 1957. Seen near Stevenage working a 'up' goods train is ex-LNER Class A4 4-6-2 No. 60028 *Walter K. Whigham*. Entering service from Doncaster Works during 1937 and named *Sea Eagle*, her name was changed to honour the then Deputy Chairman of the LNER in 1947. Allocated to King's Cross shed at the time of this photograph, she would be withdrawn from service during 1962.

Opposite top: Sunday 7 September 1957. Class A1 4-6-2 No. 60122 *Curlew* has just passed Stevenage station at the head of an 'up' express. Entering service from Doncaster Works in 1948, she would be allocated to King's Cross shed and later spend time allocated to Grantham and Copley Hill in Leeds, being withdrawn from service in 1962 whilst allocated to Doncaster shed.

Opposite bottom: Sunday 7 September 1957. Photographed from the same position as the previous image, ex-LNER three-cylinder Class V2 2-6-2 No. 60841 is also working an 'up' express. Constructed at Darlington Works during 1938, she would be rebuilt in 1959 with three separate cylinders and be withdrawn from service during 1963.

Above: Sunday 17 August 1958. The final development of the Edward Thompson 'Standard' Pacific for the LNER was his Class A2/3, of which fifteen examples were constructed at Doncaster Works during 1946 and 1947. Seen here passing the former York Racecourse platform at the head of an 'up' express is No. 60522 *Straight Deal*. Entering service during 1947, she was allocated to English sheds for most of her working life, moving to Scottish sheds in 1962 and being withdrawn from service in 1965. She had been named after the racehorse that won the 1943 Derby.

Opposite top: Sunday 17 August 1958. Passing York Racecourse platform is another 'up' express, in this instance hauled by ex-LNER Class A3 4-6-2 No. 60048 *Doncaster*. Constructed at Doncaster Works during 1924 as a Class A1 locomotive, she was rebuilt as a Class A3 in 1946, would acquire a double chimney during 1959 and be fitted with trough-style smoke deflectors in 1961. Named after the racehorse that won the 1873 Derby, she would be withdrawn during 1963.

Opposite bottom: Sunday 17 August 1958. The final member of the Arthur Peppercorn-designed Class A2 Pacific is seen here at the head of an 'up' express passing the York Racecourse platform. No. 60539 *Bronzino* entered service during August 1948 from Doncaster Works and was named after the racehorse that had won the 1910 Doncaster Cup. She would be withdrawn during 1962.

Above: Sunday 17 August 1958. Another express seen passing the York Racecourse platform is in the charge of ex-NER three-cylinder Class S3 (LNER Class B16) 4-6-0 No. 61448. Originally constructed at Darlington Works in 1923 incorporating Stephenson valve gear, she was rebuilt during 1944 with Walschaerts valve gear and would be withdrawn from service in 1964.

Opposite top: Sunday 17 August 1958. And yet another 'up' express passing the Racecourse platform at York is in the hands of ex-LNER Class D49 4-4-0 No. 62762 *The Fernie.* Constructed at Darlington Works during 1934 as a part 2 version of the class utilising Lentz rotary cam-operated poppet valves, she would be withdrawn from service in 1960.

Opposite bottom: Sunday 17 August 1958. Ex-LNER Class A4 4-6-2 No. 60007 *Sir Nigel Gresley* is also seen passing the former York Racecourse platform in charge of an 'up' express. Entering service from Doncaster Works during 1937 and named after her designer, she was based at King's Cross and Grantham sheds for much of her working life, but would be withdrawn from service in 1966 whilst allocated to Aberdeen Ferryhill shed. Purchased by a preservation trust, the locomotive has worked many main line rail tours and is currently undergoing a complete overhaul.

Above: Sunday 17 August 1958. And finally on this day at the same site as the previous photograph, Class A1 4-6-2 No. 60158 *Aberdonian* is in charge of an 'up' express. Entering service from Doncaster Works in 1949, she was allocated to King's Cross shed and was successively allocated to Grantham, Copley Hill, back to Grantham and King's Cross, before ending her days based at Doncaster and being withdrawn during 1964.

Saturday 30 August 1958. The Vincent Raven design of three-cylinder mixed traffic Class S3 4-6-0 for the NER consisted of seventy examples, which were designated Class B16 with the LNER. A comparison is here seen at Annesley shed.

Opposite top: No. 61452 in unrebuilt condition retaining her Stephenson valve gear. Constructed at Darlington Works in 1923, she would be withdrawn during 1961.

Opposite bottom: No. 61476 seen as rebuilt with Walschaerts valve gear. Having entered service from Darlington Works during 1920 and rebuilt in 1945, she would be withdrawn from service in 1963.

Saturday 30 August 1958. Designed by Henry Ivatt and introduced during 1911, the GNR Class J22 0-6-0s fitted with superheating boilers were so well received that construction continued until late in 1922, just prior to the grouping. Designated Class J6 with the LNER, a total of 110 examples saw service through to British Railway days. Seen here at Annesley shed is No. 64235, which entered service from Doncaster Works in 1914 and numbered 586 by the GNR. Becoming No. 3586 and later No. 4235 with the LNER, she would be withdrawn after forty-five years' service in 1959.

Saturday 30 August 1958. The John Robinson design of Class 9N 4-6-2 tanks introduced in 1911 for the GCR were destined to work the heavy suburban traffic out of Marylebone for many years. Seen here at Annesley shed is No. 69825, a product of Gorton Works post-grouping in 1923. She would be numbered 5046 by the GCR, later becoming No. 9825 with the LNER. She would be withdrawn during 1959. Note the missing front buffer.